TH**E MEDIA MANIA**

HEALING THE HUMAN SOUL
FROM THE DIGITAL PANDEMIC

BY SCOTT RITSEMA

The Cure for Media Mania: Healing the
Human Soul from the Digital Pandemic
By Scott Ritsema

Published by Amazing Facts International
P.O. Box 1058, Roseville, CA 95678
916-434-3880 | amazingfacts.org
afbookstore.com

Edited by Dolly Tygret
Cover by Haley Trimmer
Layout by Jacob McBlane

978-1-952505-79-9

THE **CURE** FOR MEDIA MANIA

HEALING THE HUMAN SOUL FROM THE DIGITAL PANDEMIC

BY SCOTT RITSEMA

CHAPTER 1
HOW TO BE HUMAN AGAIN 11

CHAPTER 2
WHERE IT ALL BEGAN 27

CHAPTER 3
THE DISCONNECTED CHILDHOOD 33

CHAPTER 4
ANTISOCIAL MEDIA 49

CHAPTER 5
DIGITAL PHARMAKEIA 63

CHAPTER 6
PEOPLE OF THE BOOK
IN THE AGE OF THE APP 77

FOREWORD

We live in a challenging time when it comes to Bible literacy. Fewer and fewer people, even Christians, are finding the time to dive into God's Word and become deep thinkers who integrate truth into their lives. A 2019 survey by the American Bible Society revealed that only 35 percent of adults read the Bible weekly, compared to 45 percent in 2013. That kind of trend leads to troubling consequences.

Furthermore, many of us no longer seek to be out in God's second book: nature. The Creator's plan in Genesis was for us to be in the outdoors, to garden, and to be active stewards of His creation—not creatures caged in concrete metropolitan zoos. A 2018 study by the Nature Conservancy found that only 21 percent of Americans regularly participate in outdoor activities.

I believe these negative trends can be partially explained by our culture's addiction to smartphones, mindless entertainment, and social media. It is a very real crisis that will lead to tragic cultural consequences if we don't confront it soon. We'll see less genuine human connection, less love for our neighbors and for God, and more division and selfishness everywhere. The headlines are filled with the ubiquitous and harmful impact of TikTok, Twitter, and Instagram.

So I'm thrilled that my friend, Scott Ritsema of Belt of Truth Ministries, agreed to write this brief but powerful book that exposes the harm constant media consumption is having on our present generation. Not only is Scott a great speaker for our times, but he is also an insightful writer who knows how to boldly say what needs to be said to our youth—and their parents.

I pray that you are blessed and mobilized by this book, that you will be drawn closer to God because of the advice you find herein. I also hope that you will share it in your home, church, and community.

Pastor Doug Batchelor
President, Amazing Facts International

INTRODUCTION

As the speaker and director of Belt of Truth Ministries since 2012, I have been delighted to partner with the publisher of this book, Amazing Facts International, on numerous occasions. Indeed, Amazing Facts' Study Guides and its *Bible Answers Live* radio program were pivotal in my personal journey of coming to a knowledge of the truth!

What's the story behind this book? In 2012, our culture was more than a decade into the high-speed internet revolution, a generation or two into the massively intensifying video game craze, and several generations into an obsession with entertainment—Hollywood, the music industry, televised sports, etc. And, lest we forget, it was also the beginning of the smartphone craze.

It was clear even to a casual observer of social trends that we were seeing the catastrophic impact of harmful media. More varieties of highly stimulating media used at more frequent intervals and for longer periods of time at earlier and earlier ages were bearing their fruit. This was downright alarming to people of conscience from all walks of life.

Thus, in 2012, the *Media on the Brain* seminar and Belt of Truth Ministries were born, launching me on a journey of hundreds of speaking engagements and satellite TV outreach events, including partnerships with Amazing Facts.

By 2016 it was clear that the growth curve of the media problem was rising at an exponential rate. With the percentage of Americans spending five hours or more per day on their screens, it had doubled in just four years (2012 to 2016). In response, Belt of Truth Ministries produced *The Media Mind* and *Technocracy* seminars with the prayer that it would slow down and even reverse this tide.

I'm glad that you want to be a part of this movement from God, a movement that is calling His people to a higher standard in media choices and is leading individuals from all walks of life into the joy of *freedom* from media habits that hold us captive—habits that harm our mental and physical health, shorten our attention spans, rob us of that precious commodity known as *time*, and damage the relationships we have with loved ones. In short, our Lord Jesus is at work in countless lives, working His plan of redemption, liberation, and restoration.

When this journey began for me in 2012, I had been a classroom teacher for nine years. Over the course of that decade, I saw the digital revolution take over my students' lives. Even in Christian schools, families slowly began accepting more and more worldly content from the entertainment media. In an effort to help my students become aware of the spiritual battle being waged for their minds and for their very souls, I began—with the help of Little Light Ministry and Shepherd's Call Ministry—sharing information about media with them. This led to speaking invitations to churches, camp meetings, and conferences to share these urgent messages with a generation drowning in media and entertainment.

So that's the backstory. As we begin, you'll notice that I have refrained from using academic-level citations to create a more casual reading experience. However, for those who would like a deeper dive into the subject,

I have included a bibliography at the back of this book—a list of sources that contain hundreds of citations upon which this book heavily depends.

When you're finished with this book, you are invited to enjoy what I believe to be good, quality media on many of the themes outlined herein. These can be found on the "Belt of Truth—Scott Ritsema" YouTube Channel. And be sure to visit amazingfacts.org to view my two series on Bible prophecy, *The Seven Deadly Myths in Christianity* and *America's 11th Hour: 400 Years of Providence and Prophecy.* These series provide the antidote to worldly media: Bible study!

And now, pray as you read. Think, reason, and search the Scriptures. And hear and heed the voice of God, which brings inspiration, conviction, and hope.

Scott Ritsema
Belt of Truth Ministries

CHAPTER 1
HOW TO BE HUMAN AGAIN

The average young person racks up 10,000 hours of video gaming by the age of 21, with 5 million gamers playing over 40 hours per week. Today, the average child views more than 200,000 acts of violence before the age of 18, tens of thousands of scenes of a sexual nature, and thousands of commercials for alcohol.

The average teen spends nine hours per day on entertainment and social media, with their parents not far behind at just under eight hours. At one point in 2018, Nielsen Media even reported that the media time for the average American was exactly 666 minutes per day!

Perhaps most tragically, even before the smartphone took the world by storm, American children were spending more hours watching television by the age of six than the total hours spent in conversation with their own fathers throughout their entire lifetime!

These facts and more are the reason why researchers, educators, parents, and even industry insiders began raising the alarm about the scourge of excessive media use in the digital age. But this was only the beginning.

Think about this for a moment: The average American touches, taps, or swipes their smartphone 2,617 times per day. That equates to nearly a million individual touches, taps, and swipes every year. Do we touch real things—pets, plants, people, cooking implements—a million times?

It doesn't take a research study to realize that humanity has become more engrossed in screens and more immersed in the virtual world in the 21st century. For many, social media has become their social life. Entertainment has replaced family time. And the counterfeit reality of the online world has supplanted nature, labor, books, service, and family.

In other words, the vast, varied, and beautiful tapestry of what was once the human experience has largely been discarded in favor of a screen no bigger than a few inches.

It goes without saying that screens can be used as a tool in service to God and even to enhance our humanity. But are we using the technology—or has the technology begun to use us? (More on that in a later chapter.) Is it time to learn how to be human again? "God created man in His own image; in the image of God He created him; male and female He created them" (Genesis 1:27). What does it mean that we were made by the Creator God in His own image?

Being Human Socially

Did you know that the more online-only "friends" a person has, the lonelier they are? And what the big social media titans have bragged about—that they've created the most "socially connected" generation in history—is contradicted by the fact that this is the loneliest generation ever studied. The U.S. surgeon general even used the word "epidemic" to describe this phenomenon. (We'll look more at antisocial media in a later chapter.)

A generation that is socially fulfilled, as God designed for us to be, would not have the need for a chair that hugs you back—yes, that was actually invented—or a cultural craze called the cuddling-with-strangers trend. Even if only a tiny percentage of people feel the need to pay a professional cuddler to snuggle with them so that they feel loved, it is still a sign of a serious cultural deficit.

Time spent online is correlated strongly with loneliness. Some have speculated that's merely because people who are already lonely gravitate toward social media; therefore, it's not the social media that is causing the loneliness. This hypothesis was put to the test by taking young adults off of social media completely or reducing it to 30 minutes per day. Studies on both scenarios found a decrease in loneliness. Specifically, when young adults eliminated social media for a week, a 36 percent drop in feelings of loneliness occurred!

How does our social life reflect what it means to be human?

Think about this for a moment: "God *is* love" (1 John 4:8, emphasis mine). There are three persons in the Godhead: God the Father, God the Son, and God the Holy Spirit. In Genesis 1:27, the Hebrew word *Elohim* is a plural noun translated as "God," indicating a multiplicity in the Godhead, which created humanity in His own image: male and female. The male and female become one through marriage. Then they become fruitful and multiply by having children in their own image. According to God's design, the tightest, most beautiful reflection of the image of God's love is found in the family.

Did you know that even before sin entered the picture, God identified something as being "not good"? It wasn't anything God had done wrong. It only meant that creation wasn't done yet. God had created animals in pairs. But for humans, there was just one man. Adam was alone for a time on the sixth day of creation.

God then said, "It is not good that man should be alone" (Genesis 2:18). This verse tells us something about humanity: We are not designed to be alone.

Satan enters the story after Eve was created. His aim has always been to distort what it means to be human. He succeeded in tempting Eve to disobey God when she wandered away from Adam. Satan then led Adam to blame Eve and even God for his disobedient choice. This created division in their relationships with one another and with God.

This sinful mess in the once-perfect Eden can be boiled down to broken relationships. In a subsequent chapter, we will study how prophecy indicates that the biblical family unit will be attacked in the last days and how media is fulfilling that prediction. But we'll also see how God is restoring, repairing, and renewing such relationships.

I have much more to say about the social and family implications of media, but let's continue to explore what it means to be human ...

Being Human Physiologically

Our very physiology is also worth considering.

God made mankind upright. We didn't evolve over millions of years from hunched-over primates. But have you noticed your posture while using media for many hours a day? Not only are we hunching over, but physical therapists are warning that people are reshaping the top of their spines to be permanently more forward!

The *Washington Post* also published an alarming story reporting that "horns," or small bony cranial protrusions formed from consistent bad posture, are growing on the back of young people's skulls due to overuse of smartphones.

Posture might not be thought of as the most important aspect of being human, but when our posture is poor, our respiration is shallow, which reduces oxygen to the blood and the brain. The digestive system is also hampered because we're scrunching our digestive organs. Studies have even shown that confidence, mood, and cognitive performance all suffer when we're sitting or standing with a slouched posture as opposed to an upright one.

Maybe we need to go back to this traditional counsel given to schoolchildren over a hundred years ago:

> Among the first things to be aimed at should be a correct position, both in sitting and in standing. God made man upright, and He desires him to possess not only the physical but the mental and moral benefit, the grace and dignity and self-possession, the courage and self-reliance, which an erect bearing so greatly tends to promote. Let the teacher give instruction on this point by example and by precept. Show what a correct position is, and insist that it shall be maintained (Ellen G. White, *Child Guidance*).

The physiological impacts of media use go beyond posture to actual accidents and real injuries. On YouTube, an endless array of smartphone-induced bloopers reveal people walking into things, tripping, and even falling into manholes because their noses are buried in their phones; they are oblivious to everything around them. Municipal governments have even considered putting pedestrian traffic signals *on the ground* because so many pedestrians walk into oncoming traffic! How to be human again is a relevant topic indeed.

Between 2014 and 2018, the percentage of teens who admit that they are "almost constantly" on their devices about doubled from 25 percent to nearly 50 percent.

That's a real problem when considering that some of the most important physiological impacts of media use involve sleep. God wants to bless us with good sleep every night:

- "He [God] gives His beloved sleep" (Psalm 127:2).
- "You will lie down and your sleep will be sweet." (Proverbs 3:24).

That sounds nice, doesn't it? However, media use has been shown to cause sleep disturbances. Even the casual observer is keenly aware that childhood, teen, and adult sleep deprivation is through the roof. I'll refrain from recounting the statistics on this front. The bigger question is: What is causing the sleep deprivation?

Medical doctors and scientists have shown the importance of avoiding blue-light exposure in the evening in order to optimize sleep. What is the brightest, most frequent, and most pervasive source of blue light in the evenings? It's the hours of screen time that we end our day with.

Many have tried to cheat this by using a blue-light-blocking filter on their screens or by wearing blue-light-blocking goggles, which do help a little. However, much of the sleep deprivation comes not just from the light but also from the hyper-stimulating nature of the media itself. Entertainment media, social media, and video games tend to illicit a stress-hormone response—such as the fight-or-flight hormone cortisol—that disrupts sleep quality and duration. Yes, block that blue light, but also choose low-stimulating, nature-paced media in the evenings, or no media at all to improve your sleep quality.

Being Human Thinkers and Doers

God is the Creator. We are made in His image. We can also create in our own sphere of existence. We can think and do. We have the ability to invent, compose, speak, and write. God has given us dominion over this world by serving as its stewards. Therefore, we are to be thinkers and not mere reflectors of others' thoughts.

Is today's media helping us establish our individuality? Think about this: *Time* magazine pointed out a few years ago that the average attention span for a citizen of the Western world is a mere eight seconds, which is shorter than the nine-second attention span of a goldfish!

Researchers who study creativity estimate that the average person—after mulling over an idea, pondering, weighing options, testing, and trying—typically has their "lightbulb" moment of problem solving after a 15- to 20-minute period of intense focus. That's about 1,000 seconds. And what's our attention span again? Just eight seconds. At that rate, we will never get to that moment of creativity!

Today's media is massively shortening our attention spans. We're not talking only about ADHD diagnoses but also our exposure to fast-paced media. Our brains are rewiring to expect that continual level of high-stimulant input. When real life doesn't deliver that level of stimulation, we become bored and seek something else to stimulate us. For example, 77 percent of people admit that when nothing is immediately occupying their attention, they reflexively grab their smartphones to see what's going on. This psychological phenomenon is called "stimulus-response." We have become responsive and reactive. As our individuality wanes, we become prey to the directives of the big media.

Not long ago, a friend of mine sent me an article from CNN about media. It claimed that kids today are not consuming any more media than they did in the 1980s. However, the research of Jean Twenge, a San Diego State University researcher on teens and media use, proves this wrong. Her research has found that the media use for teens has nearly tripled from the 1980s to 2016. The unsuspecting public reads the news and thinks they can always trust the information. But in this case, it was contrary to research. As Christians, we need to think for ourselves and not be a mere reflector of other's thoughts or opinions. The Lord wants "us [to] reason together" (Isaiah 1:18). He wants us to be thinkers.

Edward Bernays, the founder of modern public relations and also the nephew of famed psychoanalyst Sigmund Freud, laid the groundwork for the biggest propaganda machine in history: the advertising industry of the 20th century. In his book *Propaganda*, Bernays described the world that was already emerging at that time under his guidance:

> The conscious and intelligent manipulation of the organized habits and opinions of the masses is an important element in democratic society. Those who manipulate this unseen mechanism of society constitute an invisible government, which is the true ruling power of our country. We are governed,

our minds molded, our tastes formed, our ideas suggested, largely by men we have never heard of.

He even asked this chilling question to his fellow media masters:

If we understand the mechanism and motives of the group mind, is it not possible to control and regiment the masses according to our will without their knowing about it?

The area of our brain responsible for critical thinking, attention span, and self-control is called the pre-frontal cortex. Psychologists refer to this function as "executive attention." And tragically, especially in children whose brains are still in the developmental stages, this critical area of the brain becomes maldeveloped when exposed to excessive or over-stimulating media. It also causes people to behave more like machines than thinking humans. Moreover, our limbic system circuitry—the area of the brain for fear, aggression, lust, impulses, anxiety, anger, irritability, negativity, and aggression—becomes enhanced with the use of entertainment media.

Transformed by the Renewing of Our Minds

In addition to the developmental harms already mentioned, we will explore the tremendous mental health consequences of today's media in a later chapter. But let's get back to how to be human again and how God can restore His image in us. This restoration is the essence of our redemption.

Romans 12:2 calls it being transformed by the renewing of our minds. Paul warns us to "do not be conformed." We must chart a different course. Rather than fitting the mold of the culture around us, we choose Christ. Once we make this decision, God promises us a renewed mind and a transformed brain. So do not despair when we see so much harm from media. God told Adam and Eve to enjoy the garden of Eden and its many trees (Genesis 1:28; 2:16); likewise, we have many things to do with our time other than media. And when choosing to watch media, we can make good media choices that can be used for a healthy, balanced life.

What happens to our mental health when we abstain from media ? Dr. Victoria Dunckley, author of *Reset Your Child's Brain*, puts her patients on a program to reverse mental-health disruptions. She has successfully treated hundreds of youth with disorders ranging from depression to ADHD and anxiety to disruptive mood disorders. Parents take their children to Dr. Dunckley because she tries different treatment methods other than merely pushing pharmaceuticals.

Specifically, Dr. Dunckley puts her patients on a strict three- or four-week media fast—literally zero screen time for this period. And the results should rock the mental health profession: 80 percent of her patients had the majority of their symptoms disappear. And half of her patients had all of their symptoms disappear—in just three to four weeks of no media! In *Reset Your Child's Brain*, she explains:

> I began prescribing video game restriction ... with startling results. ... I observed how the vast majority of children exhibited certain symptoms surrounding game play—symptoms strikingly similar to amphetamine exposure—that resolved in the days or weeks of complete abstention. ... Their development would grow by leaps and bounds when screens were most restricted.

This should be front page news! Praise God! We found not just the cause of the majority of the mental health disruptions in our youth today, but also the solution! These kids have all their symptoms gone. Let's call it having a renewed mind—a transformed brain.

This gift is for each of us. Jesus has made full and free provision for forgiveness and restoration. It is the image of God being restored in humanity—redemption through faith in Him and victory over bad habits through the power of His name. It is complete freedom! In seeking Him over human-made media, we can rediscover how to be human and how to be made in His image.

CHAPTER 2
WHERE IT ALL BEGAN

When was the first "text message" sent? No, it was not on a smartphone. Think back to a more primitive time. Think back to when the world's first form of electronic media was invented. It is found further back than the television in 1927, the radio in 1906, and the silver screen in 1895. And even before Thomas Edison's phonograph was invented in 1877.

Yep. The answer is found all the way back in the mid-19th century. It was in 1844, the year Samuel Morse invented the telegraph. He tapped on a wire using the code that bears his name, thus sending what we could call a "text message." The world's first text message was sent in 1844!

If you're a student of Bible prophecy, you immediately sat up and took notice! Of all possible years, this momentous event happened to fall during the biblically prophetic year of 1844, a time prophesied about in Daniel 8:14. (Be sure to watch the Amazing Facts production *The 7 Deadly Myths in Christianity* to understand the significance of Daniel 8:14 and the year 1844.)

To be clear, the prophet Daniel was referring to the cleansing of the heavenly sanctuary and not the birth of modern media. But it is probably not a coincidence that the very year that the prophetic "end times" began was also the year the telegraph changed the course of history. It hastened the speed at which the gospel could go to every nation.

The telegraph eventually gave way to the phonograph, the radio, the television, and the internet. Now people in every nation on Earth can watch an Amazing Facts Bible prophecy seminar to learn the precious, timely truths of the book of Revelation! For example, Pastor Doug Batchelor uses this technology to appear on TVs and smartphones all at once—millions of people! And it all began in 1844.

Interestingly, Morse didn't want all the glory for his invention. His first message across the wire was quoting the Bible. He messaged,

"What hath God wrought?" (Numbers 23:23 KJV). Morse likely didn't know just how magnificently God would use his media invention in the decades that followed nor about the assortment of technologies that would flow from this historic moment to spread the gospel.

This might be a surprising thing to say when discussing the ills of the media mania in our world. After all, the telegraph story is one that seems to endorse—or even baptize— media as something of God! Well, let's think about this …

When Satan tempted Jesus three times, he quoted Scripture. Satan knows the Bible. There is no doubt that he hates it, but he is not ignorant of Bible prophecy, including the significance of 1844.

Satan distorts every good thing from God—counterfeits it, co-opts it, using it for his diabolical purposes. Music, sexuality, food, or anything else we could name, He will manipulate and distort it. It's no different with electronic media. In his deadly deceptions, Satan means business, especially after 1844, the final historic time that Bible prophecy specifically identifies. Satan knows his time is short (Revelation 12:12). And as is already evident, media has become his most efficient trap to ensnare the souls of mankind.

But as we "have no fellowship with the unfruitful works of darkness" (Ephesians 5:11), let us always keep a balanced perspective on media. It goes without saying, especially given the prophetic implications outlined above, that media can be used for God's service or by the enemy of God. Media can be used to promote truth *or* to deceive. Media can be used to edify us *or* to merely entertain us—and degrade us in the process.

How can we best make God-glorifying media choices? Philippians 4:8 is clear on what our standards as Christian should be: We behold that which is true, noble, right, lovely, admirable, excellent, and praise-worthy. But assuming that we are converted, we are to love the Father and not the world (1 John 2:15). We are to eschew worldly media since content is not the only thing that we must consider. Many questions remain:

- What kinds of media are acceptable?
- At what age is it appropriate to introduce a child to media?
- During what times of the day, at what frequency, and for what duration are best?

- What about mobile versus stationary media? Interactive screen time versus passive viewing? Are there differences?
- What about nature-paced or artificially sped up media?
- Are there certain personality traits or psychological make-ups that make certain media more acutely harmful?

In exploring all of these questions, may each of us, by the grace of God and through the power of Jesus, reclaim the human soul in the emerging digital dark age.

As you evaluate your own media use, consider doing a fast from certain types of media—or eliminate the harmful ones altogether. Abstain from certain types of media, or perhaps do a 10-day trial like Daniel and his friends did with their diet. This fast can be likened to getting your eyes checked at the optometrist's office. Maybe we all think we're seeing pretty well at the moment until we get the results. The optometrist asks you to read the third line down from the top, and you see nothing but a blur. So you reply, "Is that a G or a C or an O or a Q? I can't read that line. Am I supposed to be able to read it?"

She says, "Which one is clearer here now, option one ... (click) ... or two?"

"Um ... two," you reply.

"Which one is clearer now? Two ... (click) ... or three?"

You say, "Um ... two is clearer, I think."

"How about two ... (click) ... or four?" she asks.

"Doc, two is the clearest. Are you suggesting something here?"

Then she breaks it to you: Option one is your natural vision, but two is a prescription that does, after all, improve your vision! You thought you were seeing clearly until the exam. Then you found out there's something even better in store. Only *now* do you realize how blurry it was before! Now you have a comparison, and it is undeniable: Your vision was blurry.

Let's break down the analogy here:

- Option one is the somewhat blurry life you're living now, but which you don't realize until you see there's a better way.
- So why not give option two a try? What kind of restrictions might be worth trying? What do you have to lose? God promises that He will satisfy us; He will not withhold from us any good thing; at His right hand there are pleasures evermore (Psalm 84:11; 16:11).

God offers nothing short of the abundant life (John 10:10). Give His way a try.

CHAPTER 3
THE DISCONNECTED CHILDHOOD

While Americans were putting their newborns in an iPad bouncy seat, potty training their toddlers by using the iPotty—yes, that's a real thing—and allowing 96 percent of preschoolers to use mobile devices, the founders, CEOs, and engineers of Silicon Valley were doing things quite differently.

- When Steve Jobs was asked about his kids' use of the iPad, Americans were shocked to find out that he never allowed his kids to use one.

- When Bill Gates was asked about his kids' media use, he said he limited them to just 45 minutes per day—and only for productive or educational purposes.
- When Apple CEO Tim Cook was asked about his wish for the young people in his life, he said he wants his nephew to completely abstain from social media.

Are you noticing a trend here? Arguably the three biggest names in the American tech industry were either promoting abstinence or very strict usage of media for children.

When *The New York Times* explored what the engineers at the biggest Silicon Valley firms do with their kids, the journalist was shocked:

> The people who are closest to a thing are often the most wary of it. Technologists know how phones really work, and many have decided they don't want their own children anywhere near them. A wariness that has been slowly brewing is turning into a regionwide consensus: The benefits of screens as a learning tool are overblown, and the risks for addiction and stunting development seem high ("A Dark Consensus about Screens and Kids Begins to Emerge in Silicon Valley," October 26, 2018).

Of particular concern to these parents who best know this technology was interactive screen use—video games, mobile devices, and social media. Many of these parents openly admitted that they required their nannies to prohibit smartphone use for their young children. Video game designers said not only do they prohibit their children from certain games, but they themselves will not play them because they know how addictive the games are designed to be.

Alex Constantinople, the CEO of The OutCast Agency marketing firm, was clear that he prohibits screen use for his five-year-old. (He did allow his older children to watch a quality program with a parent; watching something educational together usually doesn't overstimulate developing brains and is less addictive than video games and social media.)

When it comes to educational technology, the most sought-after private school in Silicon Valley is the Waldorf School of Peninsula, a school that insists on a screen-free education until middle school. Another highly recommended school, the Seattle Waldorf School, prohibits screen time until high school. That's the main reason these wizards of digital dopamine put their kids in these schools!

One might object and say that a media blackout for kids means a disconnected childhood and such kids would never make it in the world. But wait! This is how the most high-tech people, who have the wealth and know-how, are choosing to help their own children's development. They love their children, and they know that digital media is not what's best for young minds.

Indeed, a viral clip of a chimpanzee scrolling through social media, selecting videos to watch, should be a wake-up call that it really doesn't take much intelligence to use a smartphone. If a chimpanzee can learn to do it, then our kids will be fine waiting to learn. Even the American Academy of Pediatrics, which hasn't exactly taken the most principled stand on childhood media exposure, gave their approval to delay children's use of digital devices. These tools are so intuitive to use, they said, that there is no rush to get kids involved with them.

When it was revealed that 96 percent of American toddlers and preschoolers were using mobile devices, it shocked child-development researchers. At the time, the American Academy of Pediatrics gave in and said, "Well, screen time is synonymous with *time*," and rescinded what had previously been a recommendation that toddlers abstain from media use. After much backlash from the child-development community, the academy reinstated their former recommendation of total abstinence. The World Health Organization did one better, however, by stating that the less media before age five, the better.

Is it even possible to not use media, other than maybe Facetime with grandma, before age five?

Well, what was childhood like a generation ago?

One viral YouTube video of interviews with grandparents, parents, and children revealed what three generations did for fun as kids. The grandparents and parents described outdoor fun, games, projects, building forts, sledding, picking fruit, and going fishing. How inspiring! Then the clips of today's children came in. It was heartbreaking to hear what they do for fun: video games, texting, and binge watching shows. Their parents even shed tears when confronted with what is happening to this generation.

In the early 1980s, Neil Postman published the book *The Disappearance of Childhood*. One might think that his book was hyperbole and alarmism, but he was dead right. Where did childhood go? Today's kids spend about one-third of their lives in school, one-third sleeping, and one-third using media. There is little time for anything else.

Richard Louv, in his book *The Last Child in the Woods*, mentioned a study in which the majority of toady's 11-year-olds admitted to never climbing a tree. Ever! Three-quarters of kids today spend less time outdoors than inmates. And the average child spends more time on video games alone than all outdoor activities combined. In short, these devices have become a virtual prison and a counterfeit reality for our children.

After years of research and observing these tragic developments, Dr. Nicholas Kardaras, author of *Glow Kids*, put it this way:

> [Kids raised with screens] were, almost universally, what I like to call "uninter-ested and uninteresting." Bored and boring, they lacked a natural curiosity and a sense of wonder and imagination that non-screen kids seemed to have. They didn't know—or care to know—about what was happening around them in the world. All that seemed to drive them was a perpetual need to be stimulated and entertained by their digital devices.

While all of this is going on with the kids, what about the parents? The average parent spends more than twice as much time just watching Netflix than all the quality time spent with their children. No wonder teens spend twice as much time on their smartphones than talking with their parents. Do you see the connection?

While their toddler- and preschool-aged children are using media devices, guess how much time the parents are spending with their children? A little less time than they should? Half the time? Barely any time? *The answer is zero!* In a focused in-house study, the majority of parents spend zero time with their children while they are using media.

In Luke 12:52, Jesus indicated that this would happen in our households: "Five in one house will be divided: three against two, and two against three." He is talking about divided homes, which is a sign of the end times (2 Timothy 3:1).

But don't despair. God gives us a wonderful promise in the last verse of the Old Testament. It tells us that before Christ comes, the hearts of the children and the hearts of the fathers will be turned toward each other again! (See Malachi 4:6.) This is an indication that they were pointed away from each other; otherwise, their hearts would not need to be turned back. But that's good news! Our hearts will become united as God intended!

Maybe our heavy-media homes are disconnected homes. Maybe the child using so much media has the real "disconnected childhood." Maybe Silicon Valley's example of raising unplugged children allows them to have a true connection with their parents, nature, labor, play, and the Lord Jesus Himself.

What can we do to turn the hearts of our family toward each other again? Let's examine a few of the questions we posed earlier. This book isn't meant to dictate a prescription for your media choices but to simply provide thought provoking information along with Spirit-filled inspiration so that we can intelligently seek God's will for our media use.

There is a wonderful passage written over 100 years ago from the popular book on the life of Christ entitled *The Desire of Ages.* Perhaps we should implement this in our own children's lives:

> The more quiet and simple the life of the child—the more free from artificial excitement, and the more in harmony with nature—the more favorable it is to physical and mental vigor and to spiritual strength (Ellen G. White).

Modern research has absolutely confirmed this! The biggest problem with childhood media exposure is that it is over-stimulating. This leads to shortening attention spans, an underdeveloped pre-frontal cortex, and an overactive limbic system. It also causes language delays, cognitive deficits, and perhaps most important, a reduction in emotional intelligence.

When considering what media to allow and at what age, start with these seven principles:

1. **Keep the media at the same pace as nature.** Even better, drop the media altogether and get into nature! But to whatever extent media will be allowed in a child's life, keep the stimulant level of that media very low. It shouldn't be fast-paced entertaining media, such as cartoons and movies. Even if the content is acceptable, the fast pace of such entertainment is harmful to brain development. Similarly, interactive screen time—games, online videos, social media—are even more overstimulating to children's developing brains. In general, consider moving away from entertainment toward edification, away from amusement toward recreation.

2. **Move from disconnected to connected—with each other!** God designed the family to be a cohesive unit in order to reflect His image and

character. Interestingly, in the *New York Times* article mentioned previously, secular Silicon Valley families did things differently than most American families. Bill Gates and Steve Jobs both insisted on mealtimes together without screens. Eating regular meals together must be a good place to start since the Bible states that our children will be growing like olive plants "around your table" (Psalm 128:3). This is also a good time to talk about the commandments of God. In Deuteronomy 6, we are told to teach them when we sit in our house, as we lie down and rise up, which is evening and morning worship, and as we walk by the way.

Other Silicon Valley families said they turn off the Wi-Fi router in the evening in order to spend time together as a family. When the older children were allowed to watch something, they did it together to keep media from separating the family. That's good to hear especially when families are texting each other from within the same house. Researchers say that the presence of devices "contaminates" the home environment and leads to increased tension in the household.

3. **Generally speaking, the later media is introduced, the better.** There are critical windows in children's brain development when we want to fill them with all the good stuff—love, security, and discipline. During this time, they should also be equipped with the social, spiritual, and emotional tools needed to make their lives a success. Even if children are watching positive, slow-paced media but are disconnected from their parents, it is harmful because it is an opportunity lost in which they could be having meaningful social interaction with each other. Interestingly, researchers studied kids who are not allowed to use media and found that when parents neglect their children and spend a lot of time on media themselves in their presence, the children have some of the same negative effects as if they were using the media! Researchers called this "second-hand screen time"—echoing the harms one gets from second-hand smoke.

   Ideally, the mother should be the teacher for the first eight to ten years of a child's life in order to form a strong and secure bond.

4. **Don't use media as a reward.** If we are going to use media, it should be because there is some educational, relational, productive, or spiritual value to it. If watching media for fun becomes the point, then we're sending the wrong message to our children about the proper use of media. Instead of being a useful tool, it becomes an indulgence. Media is a means to an end and not an end in itself.

5. **Stationary media is better than mobile.** Handing mobile devices over to children is a recipe for separation, not connection. Interactive screen time is over-stimulating and much more addictive. The content also becomes more difficult to monitor. As mentioned, for the first several years of the child, refrain from using media and then transition to viewing nature-paced media together.

6. **Infrequent and short duration.** Media use in a home with children should be the exception, not the rule. Consider having a set time when these tools will be used, but live life without media most of the time. Avoid binging and use sparingly.

7. **Focus on what you can do together.**
    Many families make the mistake of with-holding media to be strict in this area but then provide their children with nothing to do instead. What should we provide them? Ourselves! Have a happy life *together*. Do daily tasks together. Find recreational activities and projects to do as a family. This is what relationships are about. "Together" is a lost word in our culture. This advice also applies to families whose children have grown. Apply this advice with your spouse, your extended family, and with your church family.

    There are many other things to occupy our time with, such as playing music, doing community outreach, going on nature excursions, gardening—just name it! The word boredom was not found much in English-language liter-ature prior to the Industrial Revolution. What have we allowed to happen? Now, we are bored if we don't have the con-stant stimulation from media.

The Healing Happens Fast!

In the next chapter, we will address emotional intelligence, also known as EQ. Eye contact, deciphering facial expressions, the reading of emotions in others along with general social and conversational skills are examples of emotional intelligence.

Researchers and casual observers alike have noticed that children who have been immersed in screens lose some of their social skills or EQ that they normally would have otherwise. A study was done to determine what effect media has on the emotional intelligence of elementary school children. They took a group of Los Angeles fifth graders out of the city to a nature-based camp for five days. Half of the kids went on this outing; the other half served as the control group, staying in the city and playing their video games.

Before the week began, an emotional intelligence test was given to students in both groups, and then another test was given after the week was over. In between, the kids at the nature camp experienced five days of team-building exercises, archery, and hiking—all the good stuff you would expect at a camp but with no media.

What were the results?

After just five days of experiencing fulfilling social and recreational time in nature without media, these kids showed improvements in their emotional intelligence scores. Just five days!

Yes, we ought to lament what we're doing to this generation. It is a disconnected childhood—disconnected from parents and disconnected from nature. But let's reverse the damage! Disconnect from media and start connecting with nature. Enjoy doing chores, playing, reading books, listening to or playing music, spending time with grandparents, climbing trees, building, fixing, gardening, and cooking together—the options are endless!

God has made us so that we can be transformed by the renewing of our minds. His healing methods are powerful, and they take effect fast. But it does take sacrifice on our part. We have to give up our cherished habits and pour ourselves into this future generation. The duty and responsibility are ours. Will we do the right thing?

CHAPTER 4

ANTISOCIAL MEDIA

Numerous empirical studies show that involvement with a local church is associated with better health and life expectancy. It is no surprise that science confirms this reality since the church is the divine plan for connecting people to the family of God. It is where we find identity, purpose, friendship, and a mission. Most of all, we connect with Jesus.

Indeed, loving God and loving our neighbor as ourselves are at the very heart of God's law; love is the core principle from which the Ten Commandments are derived (Matthew 22:37–39). Any departure from this law of love brings harm to God's children. This could happen by distancing ourselves socially from one another, refraining from service and outreach, and replacing human interactions with social media, which is destined to lead to failure in life.

We saw earlier that social media increases loneliness, hence the title of this chapter. Anything that makes us more disconnected and lonelier should not be called "social."

Similarly, a report came out some years ago revealing that since the advent of social media, empathy has dropped 40 percent in young adults. Empathy means concern for other people by entering into their experiences and their feelings. Is this revelation any surprise since Bible prophecy indicates that in the last days, "the love of many will grow cold"? (Matthew 24:12).

Jean Twenge, author of the books *The Narcissism Epidemic* and *iGen*, points out that this data indicates we have the most narcissistic generation on record. Their self-oriented way of thinking is Satan's idea. In Isaiah 14, Lucifer announced, "I will ascend," and said he would place himself in the very position of God (v. 14). He was the first narcissist. Today, the average millennial will take 25,000 selfies during their lifetime. That's not to say that a selfie is evil, but isn't that an awful lot of pictures of oneself?

Twenge also reveals the following facts about how teens' lives today are being consumed with media and that the rest of life is just not happening as it did with previous generations:

- The percentage of teens who have a job has been cut in half.
- The average teen now gets less exercise than the average 60-year-old.
- They're also studying less.
- They do less volunteer work and fewer extracurricular activities.
- And they're getting their driver's licenses much later in life.

While that last item might appear to be a silver lining since teens are engaging in less trouble while out on the town, don't celebrate too soon. Another report revealed that when teens were asked if they would rather see their friends in person or just "chat" with them on their devices, two-thirds said that seeing their friends in person was not necessary, because a text messaging chat would do.

Let that sink in. When the majority of young people feel that in-person human contact can be discarded, there is a serious problem brewing. Perhaps that's why Chamath

Palihapitiya, an early executive at Facebook, publicly expressed regret for his participation in the formation of social media. He referred to social media as something that is "destroying how society works." He felt "tremendous guilt" that they created "tools that are ripping apart the social fabric of how society works."

This is not to say that a text message is evil. But what is it about in-person communication that is so much more fulfilling? To start with, remember that most human communications, particularly those on a relational level, are non-verbal. Proximity, facial expressions, hand gestures, and even volume and tone—and yes, sometimes silence—all communicate something that a text message could never get close to communicating.

Also, when we meet with people in person, the focus is on them. There's a greeting and a goodbye moment, both of which validate them and show them that they have your attention. And when you share a joy or a sorrow with somebody in person, they can empathize with your emotion at that moment. If somebody texts, "My mom was just diagnosed with cancer," and a cry-face emoji comes back an hour later, that's certainly better than nothing. But when that same concerning fact is shared in person, immediately their face shows empathy and perhaps a comforting hug and a prayer is offered to help relieve anxiety.

It isn't just social media platforms that are robbing us of human connection in favor of virtual quasi-communication. For many, video games are their chosen "social" medium. They are no longer just Pac-man gobbling up pellets; they are entire virtual worlds in which the mind, time, and energy become immersed. It goes without saying that subsuming the life into these fake worlds—a counterfeit reality—is not helping people become more engaged with others or have a fulfilling purpose in God's actual reality.

Silent Disruptions

Consider this interesting experiment that showed that even the mere passive presence of our smartphones disrupts the quality of our relationships.

They put pairs of people in a room to get to know each other. Participants were to place their phones on the table where they sat but were to *not use the phone* and engage in conversation. After making this new acquaintance, they took a survey to rank the quality of the friendship that was formed.

A second group of pairs was asked to do the same, but with one major difference—no phones present. Their phones were left in a different room. In their stead, a notebook was placed on the table.

Pair after pair was tested in both settings; survey after survey was taken. And when the results were tallied, there was a notably higher ranking in the quality of friendships formed among those who did not have phones present. Those who had their phones on the table, even though it wasn't used, rated a lower quality of bond.

It turns out, the mere presence of our phones can disrupt our relationships. Think about it: How many times has somebody been on their phone in your presence? Or maybe you've done it. Does it draw people closer together? Or does the presence of a phone cause a potential interruption?

Indeed, other research has found that individuals on a romantic date report that their devices usually interrupt their special time together. It's been called "technoference" in a relationship, and it comes as no surprise that it correlates with a lower relationship satisfaction.

And as we saw in a previous chapter, the family itself is being divided by technoference. This is especially tragic since the number-one factor for increasing the likelihood that young people will continually walk in the faith of their parents is that their spiritual truths were communicated in an atmosphere of relational warmth and connectedness.

One idea to reverse these concerning trends is to put the phone and the internet back into a defined location. We had better relationships when the internet was only on a computer in the office and the telephone was stationary because it was plugged into a wall. Create a "hook" for your smartphone (like a charging station) in the same way old-fashioned phones had to be on the hook when not used. Let the phone stay there while at home; use the phone only as needed and return to your day with the family. This reduces the interruptions in the kitchen, living room, and garden, where we are supposed to be *together*. It can also restore family communication and fun together.

According to child development experts, the biggest deficit in child development at the moment is in the realm of the social—the relational and the emotional, which includes the entire package of emotional intelligence, conversation skills, eye contact, and so on. But as we saw in the previous chapter, emotional intelligence can increase in just five days of no media, abundant time in nature, and lots of socializing. This effect is magnified even more when it is done in a secure, loving home environment. God's design for the family is the antidote to the effects of excessive media, especially from birth through early childhood— or as early as the nursing mother looking into the eyes of her newborn rather than into the glow of the smartphone or from enjoying the laughter, fun, singing, working, being in nature, reading, and playing together.

A Mental-Health Crisis of Unprecedented Proportions

Remember our earlier study that showed a 36-percent drop in loneliness among young adults when social media was set aside for a week? That same study found a 33 percent drop in depression when social media was set aside. Also, from 2010 to 2016, when social media was ramping up, there was a 60 percent increase in depression among teenagers, a 50 percent rise in attention-deficit disorders, along with higher rates of stress and anxiety. And it bears repeating Dr. Dunckley's findings: When you drop or strictly limit social media, all these conditions reverse! This is true for adolescents and the rest of us!

Nicolas Kardaras, author of the book *Glow Kids*, put it this way:

I've worked with over a thousand teens in the past 15 years and have observed that students who have been raised on a high-tech diet not only appear to struggle more with attention and focus, but also seem to suffer from an adolescent malaise that appears to be a direct byproduct of their digital immersion. Indeed, over two hundred peer-reviewed studies point to screen time correlating to increased ADHD, screen addiction, increased aggression, depression, anxiety and even psychosis.

Those who frequently experience the following feelings are even more likely to suffer from mental health disruptions due to social media:

- "Many of my friends are happier than me."
- "I often feel inferior to others."
- "Many of my friends have a better life than me."

If such thoughts and feelings tend to intrude into your brain, then you should abstain from social media altogether. This psychological profile is at higher risk to experience the worst effects of social media.

In fact, the fragility of teens, particularly young teens and especially teenage girls, has been shown to create a dangerous vulnerability to the intense pressures of social media. After holding steady for decades, after the advent of social media, teen suicide rose by 70 percent in just 10 years; among ages 10 to 14, the rate went up 133 percent; among 12- to 14-year-old girls, suicide went up three-fold!

We have a mental health crisis of unprecedented proportions on our hands.

These revelations and others led Sean Parker, the first president of Facebook, to lament with remorse and exasperation, "God only knows what it's doing to our children's brains."

Sean Parker is correct. God does know the nature of the problem, and He provides the solution. Remember how Dr. Dunckley's three-week media fast resolved the majority of mental-health symptoms in 80 percent of patients. We know that electronic media is causing the majority of these problems, and we know from the Bible what God's design is for a whole, healthy, and happy life.

The controversy between Christ and Satan is fully seen in social media. It's a diabolical effort when big tech plays upon the insecurities of the youth. Facebook was busted with leaked documents in which they bragged to potential advertisers that their algorithms can read the emotional state of their young users and identify when they're feeling worthless, which is the perfect time to pitch an ad to them.

Sean Parker confessed that the social media platforms he helped to build are meant to deliberately "exploit the vulnerabilities in human psychology." He admitted that "we understood this consciously, and we did it anyway." That's exactly what Facebook was bragging about to advertisers—they could intentionally exploit vulnerabilities in human psychology to sell stuff. It's a system of exploitation on the young, the weak, and the vulnerable.

Thus, it comes as no surprise that heavy social media users end up with greater feelings of insecurity and that these feelings skyrocketed among teens coinciding with the advent of the smartphone and social media.

This is probably the most important paragraph in this book. If you're struggling with a low sense of self-worth, remember that your worth is not based on your perception of your social media presence. Your value is based solely on the fact that your Creator God fashioned you in the womb (Psalm 139:13), knew your days beforehand (v.16), and gave you a purpose—to know and serve Him and others (1 Chronicles 28:9). What are you worth to God? Jesus offered His life in exchange for your life (John 3:16). That's how your value is measured by the One who made you! When all is sadness, stress, and darkness, put down the smartphone, get outdoors, and open your Bible—then be still and know that He is God (Psalm 46:10). God is the One who cherishes you, loves you, and feels every pain you feel (Psalm 56:8).

Satan gets people to repeat his rhetoric, "I'm the center of it all; look at me; I'm great"—i.e., narcissism. But he is also known as "the accuser" (Revelation 12:10). He delights in beating us down and making us feel that God doesn't love us and will not forgive us. What a lie! But it's no surprise coming from the father of lies (John 8:44).

What Do You Have to Lose?

When considering what kinds of media, on what devices, and what age for children to use, think about what they are exposed to when given unrestricted and private use of the internet. Think of the moral implications combined with the mental-health consequences outlined above. Maybe our society really dropped the ball when smartphones and social media accounts were handed over to our teenagers.

And for the rest of us, consider a media fast. Remember the optometrist analogy from earlier. Why not test it out? The Lord says to test Him in these things and see if He doesn't open the floodgates of blessing (Malachi

3:10). Yes! Taste and see that the Lord is good (Psalm 34:8). A 10-day fast is a sure way to determine if there are mental, spiritual, and emotional benefits. It'll be a different lens to compare with how life is going right now. What do you have to lose?

If you feel led to return to social media at the end of your fast, at least consider some boundaries. Some people have found success by just removing all non-essential notifications from their devices. Others, like the Silicon Valley engineer who invented the Facebook "like" button, have pulled off the bandage and completely removed all social media apps from their phones. Others use social media once a day on a computer at a defined time and for a defined duration. Use whatever steps you need to limit your media use. The idea is to keep social media from intruding and dominating your daily life.

CHAPTER 5
DIGITAL PHARMAKEIA

Many whistleblowers from inside big tech have warned us about the dangers of these devices. Sean Parker sounded the alarm about the mental health effects of social media, especially on children and youth. We've also seen Chamath Palihapitiya stating that smartphone social apps are helping to rip apart our culture's social fabric.

Another insider warned social media users with these stark words: "You don't realize it, but you are being programmed." Speaking of which, the man who invented the pull-down-to-refresh function on our phone apps was asked about the response this function elicits in the brains of users. He admitted that the function mimics the addictive slot machines in casinos—instead of spilling out money, phones light up with new messages.

Ivan Pavlov, a 19th century Russian psychologist, was known for his experiments of classical conditioning on canines. He trained his dogs to salivate when they heard a bell. They were repeatedly encouraged to associate the sound with food and would salivate even when no food was offered. Fast forward to the mid-1990s and instead of a bell, it was the dial-up sound followed by the dopamine-inducing announcement: "You've got mail!" Now it has evolved into a hundred other stimuli that illicit the dopamine pleasure response in our brains.

Sean Parker stated, "I feel tremendous guilt." He said Facebook was delivering dopamine hits to their user's brains and were programming people to be addicts. Even the inventor of the silly smartphone game called Flappy Bird felt guilty when he realized the millions of hours being wasted on it. He raked in tens of thousands of dollars per day; however, he couldn't look himself in the mirror and subsequently pulled the game from the app store.

Other addiction-causing techniques reveal the extent to which we are being programmed. One example is the infinite scroll where more content is revealed at the bottom of the page. Auto-play, that annoying function on YouTube where the next video starts right away, is another. Unlimited streaming and the encouragement of binge-watching are more addiction traps. The CEO of HBO, Richard Plepler, admitted, "What we're in the business of doing is building addicts, of building video addicts." The CEO of video-streaming service Netflix, Reed Hastings, famously said that one of their biggest competitors *is sleep*. And perhaps the most addictive of all social media functions is the "like" button. The man who invented that button regrets it because it has driven users to develop a ravenous hunger for instant validation.

But we must not forget that the approval of only one Being really matters: God's. His faithful followers will one day hear Him say, "Well done, good and faithful servant" (Matthew 25:21). We cannot do anything to earn "likes" from Him. He simply loved us enough to let His own Son die for us. We are not worthy of God's approval, but through faith in Jesus' shed blood, we become righteous. This gospel goes beyond the cheap feeling of validation obtained from social media; it carries with it an eternal weight of glory and eternal consequences.

Let us return to our big-tech whis-tleblowers. We can also learn from Dr. James Williams, a Google engineer who created one of the most important advertising metrics in internet's history. One day, he glanced at a screen in his office and noted a certain data point on the graphs of internet traffic. He called a colleague over and said, "Do you see this data point here? That is a million people we caused to do something they otherwise would not have done, and which was not even in their best interest." After experiencing a pang of conscience, Williams publicly identified big tech as "the largest, most standardized and most centralized form of attentional control in human history."

Then there's another Google engineer, Tristan Harris, who studied at the Persuasive Technology Laboratory at Stanford University. Here are his warnings:

- "Never before in history have the decisions of a handful of designers ... had so much impact on how millions of people around the world spend their attention."
- "All of us are jacked into this system. All of our minds can be hijacked. Our choices are not as free as we think they are."
- "A handful of people working at a handful of technology companies, through their choices, will steer what a billion people are thinking today."

This goes far beyond "persuasive technology." This is outright manipulation, control, and malfeasance. No wonder these engineers and CEOs are so careful about what technology they allow their children to be exposed to!

The God of heaven uses persuasion, reason, self-sacrifice, and love as His methods of influence. Satan, the serpent who was more subtle and crafty than "any beast of the field" (Genesis 3:1), works by deception and control (Revelation 13:14). This sounds a lot like how social media companies operate.

Harris also explained that something as simple as the color selection in the design of Facebook can manipulate our brains. When it first debuted, Facebook wasn't getting much engagement. This was at the time when its notification color was blue, which was designed to match the overall color scheme of the website. But a dramatic change happened when engineers changed the notification color to red.

Sound benign? Well, the engagement skyrocketed! Why? As Harris explained, it turns out that the brain registers the red color as an alarm signal. It's a trigger for the amygdala— the fear and anxiety circuit of our brain. The little alarm going off in our brain is enough to prompt us to click and "solve" the problem of the red notifications blaring at our eyes.

Addiction

Let's begin this section with a few anecdotes that give us a sense of our media-addiction problem:

- There's such a thing as smartphone-loss anxiety disorder: nomophobia
- The fear of losing one's phone ranks as high on the fear scale as the fear of dying in a terrorist attack. (This survey was done in London, where there have been many such attacks.)
- Those in their 20s and 30s ranked an internet connection as more important to their quality of life than even daylight and hot water.
- When asked to describe the feeling they get when they can't find their phones, 78 percent selected the strongest word: Panic!
- Researchers at the Massachusetts Institute of Technology found that losing a smartphone hits people emotionally in the same way as the death of a loved one.
- Neuro-marketing experts at Apple have identified that love circuits fire when we are thinking about and using our smartphones.

- Among moderate and heavy smartphone users, just 10 to 30 minutes without their phones increases feelings of anxiety with every passing minute.

In 2011, before the ubiquitous use of social media, researcher George Barna sought to identify the depths of the media-addiction problem in America. He took a seven-question survey that the American Psychiatric Association uses to diagnose addictions to substances and gambling and asked these same questions to the users of various electronic media. He wanted to see if their favorite media qualified as an addiction. What he found was shocking: The majority of Americans who answered the survey qualified as having an addiction.

There is no doubt that video games are highly addictive because they are designed to be so. Remember the video-game designer from a previous chapter who wouldn't even play video games himself or allow his children to? Well, another top video-game designer, Ian Bogost, admitted that video game addiction can be called "the cigarettes of this century." It's a harmful addiction. We all know it, but society remains in denial while the harm continues.

Dr. Nicholas Kardaras, author of the book *Glow Kids*, put it in these words:

- "I've worked with hundreds of heroin addicts ... and what I can say is that it's easier to treat a heroin addict than a true screen addict."

- "We now know that those iPads, smartphones, and Xboxes are a form of digital drug. Recent brain imaging research is showing that they affect the brain's frontal cortex—which controls executive functioning, including impulse control—in exactly the same way that cocaine does."

- "Both drug use and excessive screen usage actually stunts the frontal cortex and reduces the grey matter in that part of the brain. So hyper-arousing games create a double whammy. Not only are they addicting, but then addiction perpetuates itself by negatively impacting the part of the brain that can help with impulsivity and good decision making."

- "Your kid's brain on Minecraft looks like a brain on drugs. No wonder we have a hard time peeling kids from their screens and find our little ones agitated when their screen time is interrupted."

If you were around in the 1980s, you probably remember the public service announcement that used an egg and a frying pan as an object lesson for not using illicit drugs: "This is your brain on drugs." It is long overdue that we have the same level of alarm, concern, and intervention for a generation addicted to video games.

But more than that, worldly music, social media, pornography, entertainment, and spectator sports are also highly addictive. Victoria Dunckley, in her book *Reset Your Child's Brain*, put it this way:

> View electronics as a stimulant, in essence not unlike caffeine, amphetamines, or cocaine. Electronic screen device use puts the body into a state of high arousal and hyper-focus, followed by a crash. This over-stimulation of the nervous system is capable of causing a variety of chemical, hormonal, and sleep disturbances in the same way other stimulants can.

This is why Dr. Peter Whybrow, a neuroscience expert at UCLA, refers to these entertainments as "electronic cocaine" and why Chinese researchers adopted the term "digital heroin." Dr. Andrew Doan of the U.S.

Navy, a specialist in addiction recovery, called it "digital pharmakeia"—a name inspired by a word in Revelation 18:23 that refers to sorcery and drug use.

James Steyer, the founder and CEO of Common Sense Media, sounded the alarm in this way: "Tech companies are conducting a massive, real-time experiment on our kids, and, at present, no one is really holding them accountable." Protecting our children seems to be the increasingly impassioned refrain of researchers in many child-development fields. For example, Dr. Maryanne Wolf, author of *Reader, Come Home*, states:

> No self-respecting internal review board at any university would allow a researcher to do what our culture has already done with no adjudication or previous evidence—introduce a complete, quasi-addictive set of attention-compelling devices without knowing with possible side effects and ramifications for the subjects: our kids.

Similarly, Catherine Steiner-Adair, in her book *The Big Disconnect*, states:

> Talk of addiction is not hyperbole. It is a clinical reality. As adults, we may choose

to mess with our mind and gamble with our own neurology. But I have never met a caring parent who would knowingly risk his or her Child's future in this way. And yet we are handing these devices—that we use the language of addiction to describe—over to our children, who are even more vulnerable to the impact of everyday use on their developing brains.

Remember the 60 percent increase in depression that occurred after the advent of the smartphone and social media? This is the saddest factor of all when it comes to addiction. Every addiction reduces happiness and increases sadness. This happens even if one doesn't qualify for a depression diagnosis. When examining brain scans of an addict, it revealed that the pleasure receptors do not light up in the same way as they do in the brain of a person not suffering from an addiction.

This is the ultimate trick of the devil! He makes us miserable by promising constant pleasure. What a diabolical bait and switch! It's a temporary high that leaves one miserable. That's not a happy life he offers! Only Jesus knows how to give us an abundant life (John 10:10).

In her article "Has the Smartphone Destroyed a Generation" in *The Atlantic*, Jean Twenge states:

> Recent research suggests that screen time, in particular social-media use, does indeed *cause* unhappiness. ... The more they used Facebook, the unhappier they felt, but feeling unhappy did not subsequently lead to more Facebook use. ... The results could not be clearer: Teens who spend more time than average on-screen activities are more likely to be unhappy, and those who spend more time than average on non-screen activities are more likely to be happy. There's not a single exception. All screen activities are linked to less happiness, and all non-screen activities are linked to more happiness. ... If you were going to give advice for a happy adolescence based on this survey, it would be straightforward: Put down the phone, turn off the laptop, and do something—anything—that does not involve a screen.

Thanks Be to God!

The Bible announces this powerful gift and promise: "Thanks be to God, who gives us the victory through our Lord Jesus Christ!"

(1 Corinthians 15:57). There is infinite power in the name of Jesus. There is invincible victory through His Spirit.

Throw away your harmful media! Yes, do it right now. By doing this, you are taking a step of faith. Remember to claim the promises of God while taking this step. He will give you the victory through Jesus, and then you will be more than a conqueror! (Romans 8:37).

As far as our young people go, maybe we can take some wisdom from a teenage girl interviewed for the *Atlantic Magazine* article. This insightful lady didn't think that restricting media use was considered missing out. Rather, she lamented that her generation never knew life *without* the smartphone.

Does that make you pause for a moment? Have we taken away the privilege of a simple childhood from them? Have we thrust them into a harmful addiction? If we take a strong stand to protect our children from the harms of alcohol or nicotine, we are not considered "strict" or "controlling" parents. Rather, we are protecting our children *from* something that would control them! We are *enhancing* their freedom and validating their individual will so that they will not be a slave to something.

It's the same with media. Here is what the founder of a Waldorf School in Oregon says:

> We believe that computer skills in the classroom should be postponed to high school [to create] a moral foundation of freedom of choice, instead of being totally dependent on electronic media.

Notice, she's not even talking about entertainment and social media but basic computer skills! Most important, what is the reason for being so strict when it comes to media? To preserve a child's freedom of choice!

So long as we are filling children's lives with ourselves, our time, and our devotion to their happiness, don't let anyone say that protecting children from an addiction is controlling. The children will thank you when you are older. Wouldn't you? This is practicing the Golden Rule based on Matthew 7:12, "Do unto others what you would have them do unto you."

CHAPTER 6

PEOPLE OF THE BOOK IN THE AGE OF THE APP

We've already seen many antidotes to our media problem, which include family, nature, and rediscovering how to be human again. But the ultimate answer is to seek Jesus Christ in His Word.

There is nothing more calculated to energize the mind and strengthen the intellect than to study the Word of God. No other book is so potent to elevate the thoughts and to give vigor to the faculties as the ennobling truths of the Bible. If God's Word were studied as it should be, a person could develop a breadth of mind, a nobility of character, and stability of purpose that is rarely seen.

In this closing chapter, let's explore the benefits of book reading in general—and then the Bible in particular.

Virtue

Dr. Maryanne Wolf, in her book *Reader, Come Home,* identifies four benefits of reading books. As you encounter these, you'll think, *"I think I've heard this somewhere."* You have—in the Bible!

1. Critical thinking
2. Creativity
3. Personal reflection
4. Empathy

All of these are found in the Bible, especially empathy. It is the essence of the message of the Scripture: God is love. Because He first loved us, He calls us to love Him with all of our heart, soul, mind, and strength, and to love our neighbor as ourselves (Matthew 22:37–39). To truly love our neighbors as ourselves, we have to empathize with their afflictions and weaknesses as Jesus did. He intermingled love with empathy, and we must do the same.

Another benefit brought out about reading is personal reflection. Isn't that what the apostle Paul invites us to do in 2 Corinthians 13:5: "Examine yourselves as to whether you are in the faith"? The other two points are creativity and critical thinking, which are integral parts of what it means to be a human in God's image.

Secular researchers, such as Dr. Wolf, can get to the heart of God's plan for us by looking at facts objectively. The next step for them and us is to seek to glorify God in all that we do because He has been so good to us.

There is such unbounded virtue in studying God's Word. It is a privilege that humanity has not always had. For example, during the pre-literary times in ancient history, all knowledge had to be conveyed orally. Then there was the Dark Ages, a time when the Bible and other literature were withheld from the masses as a form of social control.

Still, God knew that we would need brains to turn off the media and read! Ponder these facts for a moment: There are more connections in a single cubic centimeter of the brain than there are stars in the galaxy! And when we learn to read as children, we are forming a reading circuit within the two brain hemispheres. This circuit continues to the four lobes of each hemisphere and to the five layers of the brain. All the parts of this reading circuit piece together and grasp the meaning of letters and words in a fraction of a second. We are so fearfully and wonderfully made!

God wants us to love Him with all of our mind (Luke 10:27). We usually think of love as a heart issue. Indeed, it is. But it is also a mind issue. It requires a mind that comprehends and receives a love of the truth (John 8:31–36), and then a mind that chooses to love God in return.

How can we better love God with our mind? When we become intelligent in the Scriptures, we are improving our ability to discern truth from error in this age of deception.

Literacy and Intelligence Are Under Attack

Even by the 1980s, as recorded in Neil Postman's book *Amusing Ourselves to Death*, it was shown that the more television we watch, the dumber we get. By 2003, kids were performing academically two grades below where they had been in the 1970s.

How about in the era of the smartphone? Are we getting smarter? Since the advent of the smartphone and social media, SAT scores in language and reading dropped 13 points each in just 10 years. And a $300-million study funded by the National Institutes of Health found that just two hours of screen time reduced thinking and language scores in 9- and 10-year-olds.

Are thinking skills and language skills important for the Christian? It is true that a strong intellect will not save us. The learned

Pharisees in Jesus' day are a good example of this. But at the same time, we should ask, "Who will be more easily deceived: those who are improving or those who are degrading their thinking and language abilities?" Without a doubt, Satan's last-day deceptions prophesied about in the Bible will be more potent when exercised upon a mind-numbed generation.

If you take grade school-aged kids and give half of them a video game console, in just four weeks, those kids will have:

- Significantly lower scores in reading
- Significantly lower scores in writing
- More teacher reported learning problems
- Less time spent doing homework

And believe it or not, literacy itself is under attack. By that, I don't mean people aren't able to use phonics to sound out words. I am talking about reading comprehension. This is the kind of reading we use when studying the Bible and other truth-filled, uplifting books. There is a significant difference in how we tend to read online versus how we read books. From the *Washington Post* article entitled "Serious Reading Takes a Hit from Online Scanning and Skimming, Researchers Say":

To cognitive neuroscientists [this] experience is the subject of great fascination and growing alarm. Humans, they warn, seem to be developing digital brains with new circuits for skimming through the torrent of information online. This alternative way of reading is competing with traditional deep reading circuitry developed over several millennia.

There is a literary reading style and a digital reading style, and they are completely different!

To be sure, the internet is a great resource. But the more we do the digital style of reading, the more we lose our ability to read in the traditional way—the way the Bible requires. Should the goal then be to eliminate the internet? No. But while we set aside time each day to read the Bible and other good books, let's *retain* our true, deep, literary minds. Then, when we go online for productive and edifying purposes, let's *retrain* real literacy by slowing down and thinking more deeply.

The way we read online tends to be chopped up into spasmodic bursts. It jumps from here to there and does not string together logical thoughts. This reduces our reasoning capabilities. It has been proven in studies that even reading *the very same text* on a screen versus in a book resulted in lower comprehension, memory, and critical analysis!

A good suggestion to follow is to slow down and focus when online. Follow this up by reading more books (think non-fiction). In fact, people who read books live longer! And to the contrary, the more we are online, the higher our stress hormone production will be. This toxic stress hormone exacerbates every known disease. Reading books is also neurochemically energizing, while surfing social media is neurochemically depleting.

What if, instead of spending so much time online, we began our day with a thoughtful hour contemplating the life of Christ? This can be done by reading about His final scenes and meditating on each point. Then prayerfully ask for a higher culture of mind that encompasses both the spiritual and intellectual. Research has shown that just 12 minutes of thinking about a God of love improves frontal lobe function. Religious exercises have been proven to lower stress and improve well-being on every level— social, emotional, and physical.

But this more contemplative life of study and thought is presently going down the tubes. In the 1950s, nearly all 12th graders read a book or periodical most days of the week. By 1976, that figure dropped to only 60 percent of 12th graders. And by 2016, only 15 percent of 12th graders were reading books or periodicals most days of the week. This is a dangerous social profile of a culture that is ripe for the devil's last-day deception.

Where Did My Memory Go?

Lloyd's Insurance Group recently analyzed incidents involving forgetfulness, such as bathtubs over-filling, pots left on the stove, and lost keys. They estimated that in the first decade of the smartphone, the human memory span dropped from 12 minutes to 5 minutes. That's more than a 50 percent decline in 10 years.

One survey in the United Kingdom prompted the headline in *The Telegraph*, "'Google-it' Mentality Leaves School-leavers Unprepared for University, Study Finds." They queried university admissions officers to evaluate the skills of this smartphone generation. Compared to university candidates in the past, they reported that this generation:

- Is unable to remember facts
- Is unable to think
- Is unable to manage their workloads
- Is unable to manage their time
- Has a "Google-it" mentality

Let's admit it. Far too often, all of us have said, "I don't know. Google it."

In one survey, 91 percent of respondents described their smartphones as "an extension of my brain." Think about that danger. We outsource our collection of information and facts to a soulless corporation; isn't this another way our humanity is being lost? We can no longer immediately access that prior knowledge *within the circuits of our own brains* in order to make new neural connections and develop higher meaning within the broader contexts of those facts. Isn't this paving the way for the ultimate deception?

The CEO of Google, Sundar Pichai, admitted that the minds of a generation are being harmed. He used rather neutral language when admitting this to journalist Charlie Rose, saying that big tech is "altering cognition and affecting deeper thinking." Altering? Affecting? Maybe we should say "reducing" cognition and "destroying" deeper thinking. That is the fruit of the "Google-it" mentality.

Infantile Generation

British neuroscientist Susan Greenfield famously referred to the smartphone generation as an infantile generation. This is not only because of the nonsense entertainment that was already around, or because of pop media that's been an affliction for over a generation, or because popular music lyrics are measurably dumber than just one generation ago; but also because of the vain shallowness fostered by social media. This is her biggest concern.

It's the generation of the sound bite. All that matters to many smartphone users is the momentary satisfaction they get from a dopamine hit on their media. It is the age of instant gratification. Take *Insta*gram, for example. It means *insta*-this and *insta*-that. The *insta*-culture is upon us! What about studying *history?* That's irrelevant. Or studying *Bible prophecy?* No, checking the notifications on my smartphone is more gratifying.

Moreover, we are being habituated to ever shorter and faster units of thought. On social media, background knowledge isn't brought to the forefront like it is when studying the Bible. This information enable us to make comparisons or analogies that form the groundwork for inferences or new knowledge. This isolated fact or impression leaves no wider impact on our minds, our worldviews, or our choices.

Again, if your mind is not actively engaged in the process of forming knowledge, then who is shaping your mind? Diminished attention combined with a reliance to look things up has crippled our motivation to learn things for ourselves. Such a passive relationship with knowledge diminishes our storage of knowledge and, thus, our future ability to make meaningful connections and inferences.

Increase in Knowledge—or a Digital Dark Age?

The prophet Daniel said that knowledge would increase in the last days (Daniel 12:4). Some might be inclined to see the information super-highway, the internet, as a fulfillment of that prophecy. But what is knowledge in the Bible?

> This is eternal life, that they may know You, the only true God, and Jesus Christ whom You have sent (John 17:3).

In the Bible, to "know" means an intimate acquaintance or to have a relational connectedness. We realize that we live in disconnected times and even more so when it comes to our relationship with God. Maybe the broader culture of the Information Age isn't fulfilling that prophecy of Daniel. Having access to the information on Google is not the same thing as having biblical knowledge.

Maybe a generation that is becoming more illiterate and has less empathy coupled with more narcissism than any generation before is far from fulfilling Daniel's prophecy;

it is more akin to the society described in the Revelation 13—a society in which the Dark Ages makes a comeback, a time of repression and tyranny over the individual's conscience.

Is this where we are going? This brings me to the subtitle of this book: *Reclaiming the Human Soul in the Digital Dark Age*. Prophecy does indicate a return to Dark Ages principles where individual reason and conscience are surrendered upon the altar of a final, forced, and counterfeit worship. (Learn more about that in *The Seven Deadly Myths in Christianity* at Amazing Facts.) The cloud of darkness seems to be falling upon humanity. The digital age may very well be the new Dark Ages.

It is left for us to individually seek the Light—Jesus, the light of the world. To know Him is to have eternal life. That's true knowledge. Thus, Daniel's prophecy will be fulfilled within your life and your family's. Knowledge may increase as we study His Word, seek His face, know His love, and be transformed by the renewing of our minds (Romans 12:2).

It is through Christ—the only way, the only truth, and the only life (John 14:6)—that the human soul may be reclaimed in these degrading and unprecedented times. Let Him have all the glory, honor, and power in your life. Allow Him to free you from the addictions and chains that hold the soul captive and benumb the mind. He will then restore the image of God in you. This is redemption full and free.

In the words of the famous hymn writer Charles Wesley:

> *My chains fell off, my heart was free,*
> *I rose, went forth, and followed Thee. ...*
> *Amazing love! How can it be?*
> *That Thou, my God, shouldst die for me?*

BIBLIOGRAPHY

This is a list of the sourced secular books for the facts and science references on media included in this book. The inclusion of these titles is not meant to serve as an endorsement of every sentiment expressed in them; consider their content discerningly and always through a biblical lens.

- *The Desire of Ages* by Ellen G. White
- *Child Guidance* by Ellen G. White
- *Alone Together: Why We Expect More from Technology and Less from Each Other* by Sherry Turkle
- *Deep Work: Rules for Focused Success in a Distracted World* by Cal Newport, Jeff Bottoms, et al.
- *The Last Child in the Woods* by Richard Louv

- *Reset Your Child's Brain: A Four-Week Plan to End Meltdowns, Raise Grades, and Boost Social Skills by Reversing the Effects of Electronic Screen-Time* by Dr. Victoria L. Dunckley
- *Glow Kids: How Screen Addiction Is Hijacking Our Kids—and How to Break the Trance* by Nicholas Kardaras
- *The Big Disconnect: Protecting Childhood and Family Relationships in the Digital Age* by Catherine Steiner-Adair
- *Irresistible: The Rise of Addictive Technology and the Business of Keeping Us Hooked* by Adam Alter
- *The Blessing of a Skinned Knee: Using Timeless Teachings to Raise Self-Reliant Children* by Wendy Mogel
- *iGen: The 10 Trends Shaping Today's Young People—and the Nation* by Jean M. Twenge Ph.D., Madeleine Maby, et al.
- *Reader, Come Home: The Reading Brain in a Digital World* by Maryanne Wolf
- *The Glass Cage: How Our Computers Are Changing Us* by Nicholas Carr
- *The Shallows: How the Internet Is Changing the Way We Think, Read and Remember* by Nicholas Carr
- *Hooked: How to Build Habit-Forming Products* by Nir Eyal and Ryan Hoover

www.beltoftruth.tv

All of **Scott Ritsema's** video seminars in one place! Ad-free, censorship-free...

- **Media on the Brain**
- **The Media Mind**
- **Raising the Remnant** (parenting seminar)
- **Schooled:** The deliberate agenda to destroy individuality, reduce intelligence, and re-engineer society
- **Bible prophecy**
- **Current events/current culture**
- **History**
- **Religious liberty, America's founding, the history of abortion**
- **Overcoming lust**...
- ...and much more!

SIGN UP NOW

BELT OF TRUTH
MINISTRIES

www.beltoftruth.tv

TRUTH-FILLED MAGAZINES!

Check out our colorful, biblical magazines that will open your eyes with in-depth, fast-paced teaching and thrilling, eye-catching graphics.

- **The Afterlife Mystery**
  Decoding Death, Hell, and Eternal Life

- **Amazing Health Facts!**
  8 Bible Secrets for a Longer & Stronger Life!

- **America in Bible Prophecy**

- **The Bible Truth About Hell**

- **Cosmic Conflict**
  The Origin of Evil

- **Daniel & Revelation**

- **The Day of the Lord**

- **A Divine Design**
  The Jewish Temple in History, Prophecy, and Your Life

- **Earth's Final Warning**
  The Three Angels of Revelation

- **The Final Events of Bible Prophecy**

- **Hidden Truth**
  Amazing Bible Facts Revealed!

- **Kingdoms in Time**
  History's Greatest Bible Prophecies!

- **The Rest of Your Life!**
  Everything You Need to Know About the Sabbath

Find all these and more at
afbookstore.com

Find Peace, Power, and Purpose for YOUR LIFE!

AmazingBibleStudies.com

Enroll in our FREE online Bible study course and discover:

- What happens after death
- The way to better health
- How to save your marriage
- The surprising news about hell
- Why the Bible is relevant today
- The "mark of the beast"
- Who really gets "left behind"
- ... and much more!

Or enroll in the **FREE** postal mail course! Send your name and address to:

P.O. Box 909
Roseville, CA 95678